Learning About Plant Growth with Graphic Organizers

Jonathan Kravetz

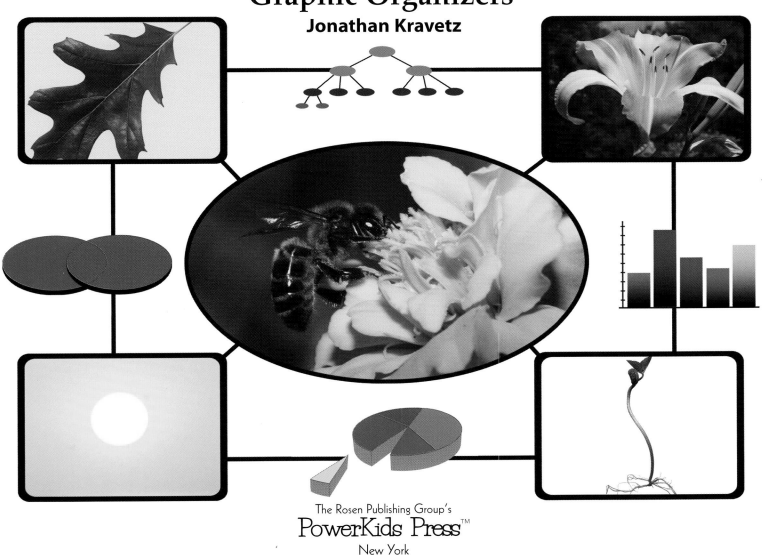

The Rosen Publishing Group's
PowerKids Press™
New York

For my friends, with appreciation for everything over the years

Published in 2007 by The Rosen Publishing Group, Inc.
29 East 21st Street, New York, NY 10010

Copyright © 2007 by The Rosen Publishing Group, Inc.

All rights reserved. No part of this book may be reproduced in any form without permission in writing from the publisher, except by a reviewer.

First Edition

Editor: Jennifer Way
Layout Design: Julio A. Gil

Photo Credits: Cover, title page (center), p. 19 © www.istockphoto.com/Arlindo Silva; cover, title page (top left, bottom right) © PhotoDisc; cover, title page (top right), p. 7 (center right) © www.istockphoto.com/Kristian Peetz; cover, title page (bottom left) © Digital Vison; cover, title page (top left) © PhotoDisc; p. 7 (left) © www.istockphoto.com/Kenn Kiser; p. 7 (center left) © www.istockphoto.com/William Walsh; p. 7 (right) © www.istockphoto.com/Greg Nicolas; p. 8 (left) © Paul A. Souders/Corbis; p. 8 (right) © David Aubrey/Corbis; p. 12 © Bojan Tezak; p. 16 (flower pot) © www.istockphoto.com/Ryan Kelly; p. 16 (bee) © www.istockphoto.com/Michael Gatewood.

Library of Congress Cataloging-in-Publication Data

Kravetz, Jonathan.
  Learning about plant growth with graphic organizers / Jonathan Kravetz.— 1st ed.
      p. cm. — (Graphic organizers in science)
  Includes index.
  ISBN 1-4042-3413-6 (lib. bdg.) — ISBN 1-4042-2208-1 (pbk.) — ISBN 1-4042-2398-3 (six pack)
  1. Growth (Plants)—Study and teaching (Elementary)—Graphic methods—Juvenile literature. 2. Plants—Development—Study and teaching (Elementary)—Graphic methods–Juvenile literature. I. Title. II. Series.
  QK731.K73 2007
  571.8'2—dc22

2005034621

Manufactured in the United States of America

Contents

What Is a Plant?	5
Types of Plants	6
Roots and Stems	9
Photosynthesis	10
Leaves	13
Plant Seeds	14
Reproduction	17
Pollination	18
Flowers	21
Plants and Us	22
Glossary	23
Index	24
Web Sites	24

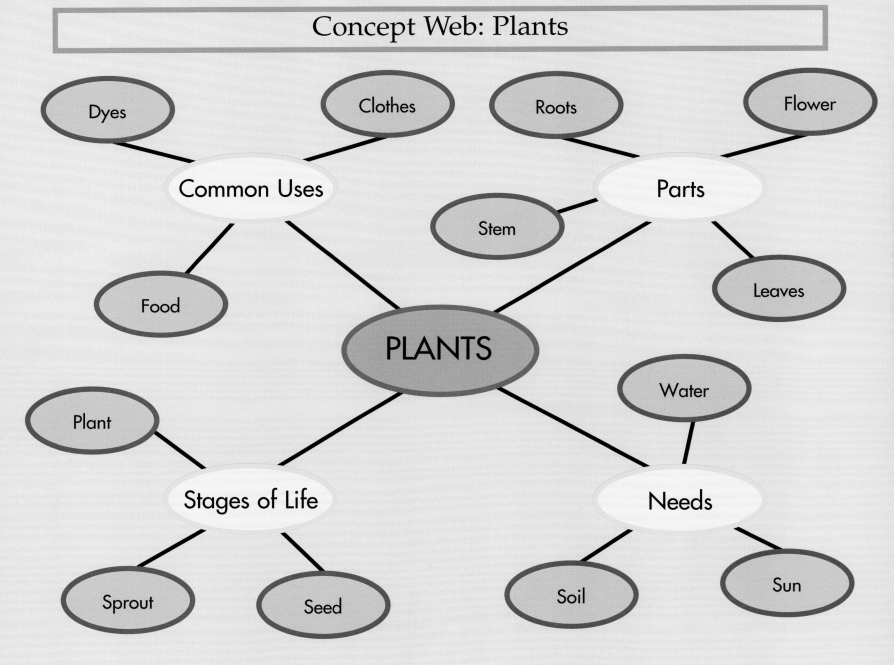

What Is a Plant?

There are around 400,000 different plants in the world, and new ones are still being discovered. Some plants are small and thin, like grass. Others are large. The giant redwoods in California are the tallest plants on Earth and can grow as high as 360 feet (110 m).

The basic parts of most plants are the root, stem, and leaf. Fruits, flowers, and seeds are also parts of certain plants. Most plants need sunlight, soil, and water to grow. Some parts of a plant, such as leaves, flowers, and fruits, usually grow to a set size. Other parts, such as roots and stems, may keep growing throughout a plant's life. Plant growth is slow but powerful. The force of the growth is so strong that some plants can push their way through sidewalks.

This graphic organizer is called a concept web. Concept webs are used to organize facts about a subject. The subject goes in the middle, and the facts go around it. This concept web helps us organize some basic facts about plants.

Types of Plants

Plants are classified, or grouped, based on whether they have a system of tubes to carry water and **nutrients** through the plant. Plants that have this system are called vascular plants. Plants that do not have this system are called nonvascular plants.

Vascular plants are classified based on whether they produce seeds. Plants that do not produce seeds include ferns. Plants that produce seeds are more common and are classified as **angiosperms** or **gymnosperms**. Angiosperms have covered seeds that grow inside a fruit. Angiosperms are flowering plants.

Gymnosperms have seeds that grow inside a cone. They do not have flowers. **Conifers**, such as pine trees, are the best-known gymnosperms. Most conifers are evergreens. That means they grow new leaves year-round.

The tree chart is a type of graphic organizer that shows how things are related to one another. Tree charts are often used to show how to group things. In this chart we group plants.

Tree Chart: Classifying Plants

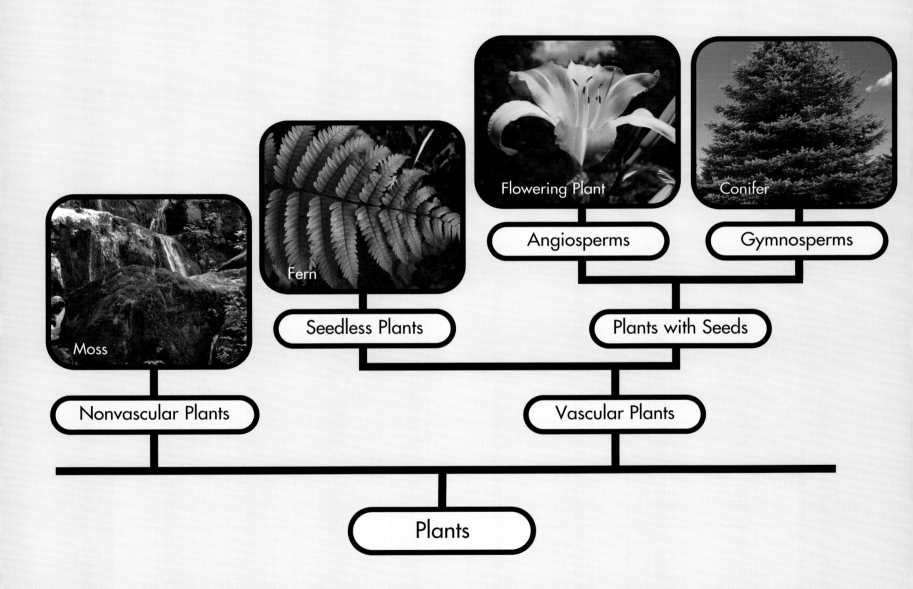

Compare/Contrast Chart: Root Systems

Taproot System

Fibrous Root System

	Examples	Root Type	Grows
Taproot System	Carrot, beet	One thick root	Root grows straight down
Fibrous Root System	Grasses, maple trees	Many roots in a bunch	Many roots spread to the side and down as they grow

Roots and Stems

Roots are the parts of plants that grow under the ground. Roots hold the plant in place and take in water and other food the plant needs to grow. Roots can also store food the plant uses to live. There are two basic types of roots. **Taproots** have one main root. Carrots and beets are taproots. Plants that have a **fibrous root system**, such as a maple tree, have many roots that branch out into the soil.

Stems are a part of the plant that you can see above ground. Stems hold up the plant. They also carry water and nutrients to and from all parts of the plant. Plants produce food in the leaves. That food moves to other parts of the plant through the stem.

A compare/contrast chart is a graphic organizer that you can use to compare things. This compare/contrast chart compares taproot systems to fibrous root systems.

Photosynthesis

All green plants need water, nutrients, sunlight, and a gas in the air called carbon dioxide to live. Plants make their own food through **photosynthesis**. Photosynthesis takes place mainly in the **cells** of a plant's leaves. Inside these cells is green matter called **chlorophyll** that absorbs, or takes in, sunlight. Chlorophyll makes plants green.

Plants draw water from the soil through their roots and get carbon dioxide from the air through the surface of their leaves. Carbon dioxide and water combine to produce glucose and oxygen. Glucose is a type of sugar that plants use for food. Oxygen is a gas that people breathe. Some of the food created by photosynthesis is used to help the plant grow. The food that is not needed right away is stored in the plant.

Sequence charts show the steps of something in order. This chart shows how photosynthesis works.

Sequence Chart: Photosynthesis

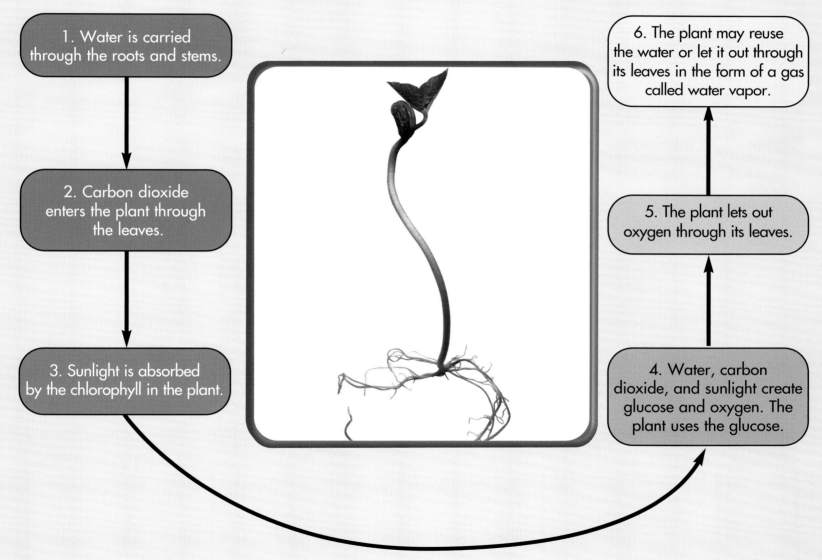

1. Water is carried through the roots and stems.

2. Carbon dioxide enters the plant through the leaves.

3. Sunlight is absorbed by the chlorophyll in the plant.

4. Water, carbon dioxide, and sunlight create glucose and oxygen. The plant uses the glucose.

5. The plant lets out oxygen through its leaves.

6. The plant may reuse the water or let it out through its leaves in the form of a gas called water vapor.

Chart: How Leaves Change Through the Seasons

Spring

Summer

Fall

Winter

	Spring	Summer	Fall	Winter
Leaves	Buds break open, and leaves or flowers begin to open.	The trees are in full leaf and are green. Many have already flowered and are producing seeds.	The leaves begin to change color and fall off the tree.	Trees have no leaves.

Leaves

Plants make food in their leaves through photosynthesis. Sunlight is necessary for photosynthesis. Leaves usually grow broad and flat so that they can more easily collect sunlight. Leaves also have many thin tubes called veins in them to carry water and food within the leaf.

There are two basic types of leaves. Simple leaves are made of a single leaf blade connected to the stem. Oak and maple trees have simple leaves. A compound leaf is a leaf made up of a bunch of smaller leaves called leaflets. Ash and locust trees have compound leaves. The surface of a leaf has a waxy coating, which keeps the plant from losing too much moisture. Leaves do not stay year-round on most trees. They bud in spring and then fall to the ground in autumn.

A chart is a simple way to organize facts. This chart explains how leaves change through the seasons.

Plant Seeds

Most plants begin life as seeds. Seeds have a tiny plant inside them with leaves, stems, and root parts. With enough water, the right weather, and the right soil, a seed begins to grow into a new plant. This is called **germination**. Only gymnosperms and angiosperms have seeds. Before seeds begin to grow, they are often carried away from the parent plant. Seeds can be spread by the wind, by animals, or by water. Before a plant begins to grow, it may remain dormant, or in a resting state, for weeks, months, or years.

In angiosperms the seeds are stored in fruit. The fruit grows around the seeds to keep them safe. People and other animals often eat fruits, such as apples and bananas. In gymnosperms the seeds are not covered, and they form on cones.

A picture chart is a graphic organizer that uses pictures to organize facts. This picture chart explains some of the differences between monocots and dicots.

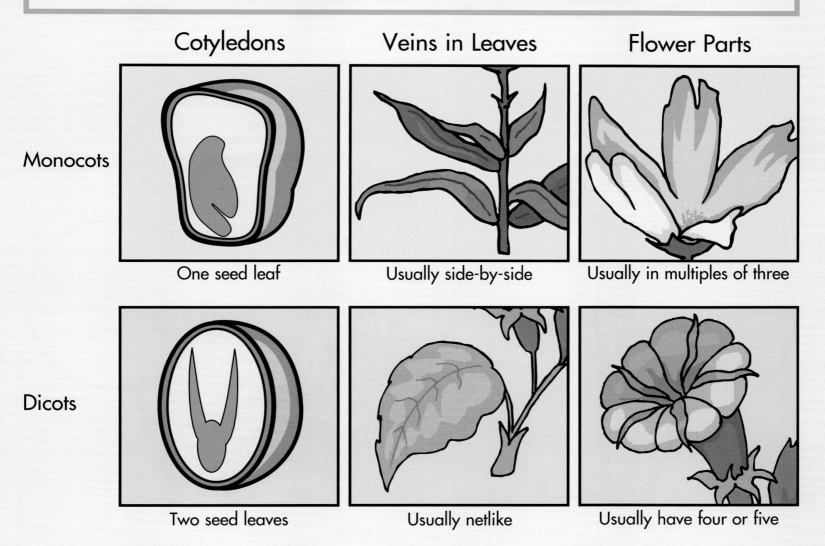

Flowering plants have either one or two cotyledons, or seed leaves. Plants with one cotyledon are called monocots. Plants with two cotyledons are called dicots.

Cycle Chart: The Life Cycle of a Flowering Plant

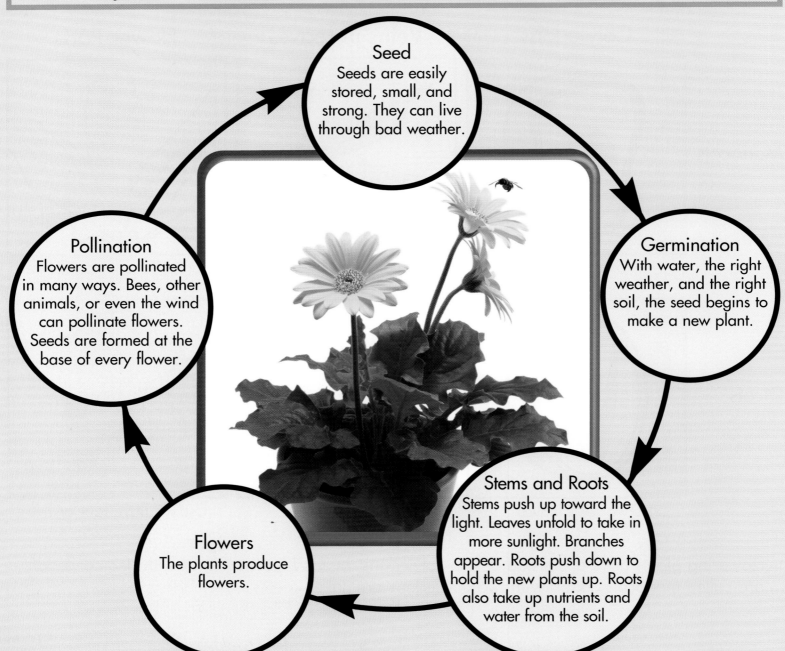

Seed
Seeds are easily stored, small, and strong. They can live through bad weather.

Germination
With water, the right weather, and the right soil, the seed begins to make a new plant.

Stems and Roots
Stems push up toward the light. Leaves unfold to take in more sunlight. Branches appear. Roots push down to hold the new plants up. Roots also take up nutrients and water from the soil.

Flowers
The plants produce flowers.

Pollination
Flowers are pollinated in many ways. Bees, other animals, or even the wind can pollinate flowers. Seeds are formed at the base of every flower.

Reproduction

Most plants **reproduce** by sexual reproduction. They begin life as seeds. The seeds are made when male and female sex cells come together. This is called **fertilization**. Angiosperms have flowers that help the plant pass male sex cells, called pollen, to the female sex cells. **Insects**, such as bees, are often needed to carry pollen. In most gymnosperms the wind carries pollen to the female part of the plant. The cone is the female part of the plant. With the right conditions, roots, stems, and leaves will start to germinate. The new plant is called a sprout.

Mosses and ferns are seedless plants. They reproduce by asexual reproduction. In one of the most common forms of asexual reproduction, plants produce **spores** that are carried by wind or water and grow into new plants.

This graphic organizer is called a cycle chart. A cycle chart shows how something grows from one form to another. This cycle chart shows a simple outline of the life of a flowering plant.

Pollination

Pollination is the spreading of pollen to the female parts of plants so that fertilization can happen. In angiosperms it is the flower that must be pollinated since it holds both the male and female sex parts of the plant. There are many ways that flowers get pollinated. Insects and other animals, such as birds, help some flowers become pollinated. They are drawn to flowers by their bright colors and scents. Insects feed on a sugary food in the flower called nectar. While feeding the insect may pick up pollen on its body. If it then goes to another flower of the same kind, the pollen can rub off on that flower. Many flowering plants in cold or wet places where there are fewer insects let the wind carry pollen. In gymnosperms pollen is usually carried by wind.

A flow chart is a group of pictures showing how something works. In this flow chart, we see how pollination works.

Flow Chart: Pollination

- An insect, such as a bee, is drawn to a flower by its bright colors and smell.
- As the bee feeds on nectar, it gets pollen stuck to its body.
- The bee flies around and is drawn to a flower of the same type of plant.
- The bee, while feeding, rubs pollen on the stigma.
- The pollinated plant produces seeds that become a new plant. The cycle starts all over again.

Diagram: Parts of a Flower

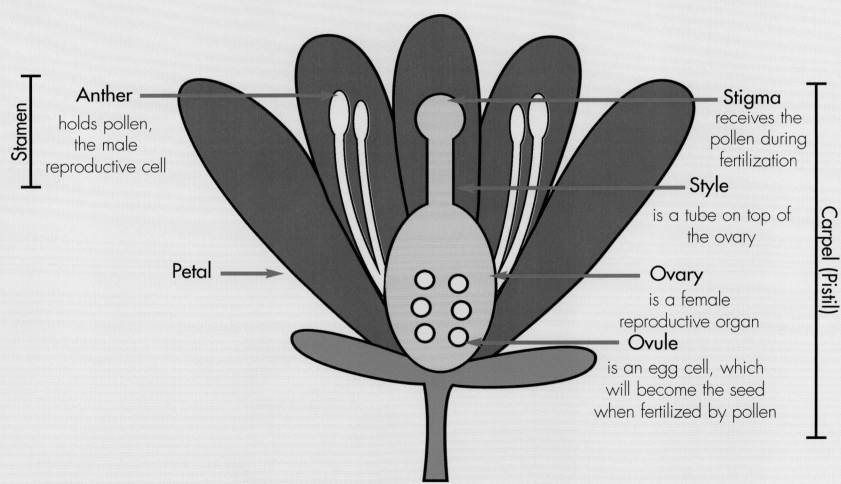

Flowers

Flowers are the parts of angiosperms that help them reproduce. The female part of a flower is called a carpel. There may be only one carpel in a flower, or there may be several joined together. The carpel or carpels joined together are known as the pistil. The carpel has three parts, the **stigma**, the **style**, and the **ovary**. The stigma is connected to the long, tubelike style. The style leads to the ovary, which holds ovules, or egg cells.

The male parts of flowers are called **stamen** and usually surround the carpel. The stamen produces pollen. Pollen is found on the **anther**. During fertilization pollen produces a tube that travels down the style. The male cells then travel down this tube and join with the ovule. The fertilized ovule becomes a seed, and the ovary becomes the fruit.

A diagram is a drawing showing the parts of something. In this diagram we see the different parts of a flower.

Plants and Us

People could not live without plants. The oxygen we breathe comes from photosynthesis and has been building up for millions of years! Plants also depend on animals and people for their lives. Animals and people breathe oxygen and give off carbon dioxide. This is called respiration. Plants use carbon dioxide for photosynthesis. The cycles of photosynthesis and respiration help keep Earth's natural balance of oxygen, carbon dioxide, and water.

Plants are also an important supply of many things for us. Wood from trees is used for buildings, furniture, and paper. Plants also provide us with food, cloth, paper, dyes, rubber, and many other things. From seed, to sprout, to full-grown adult, plants work together with people.

Glossary

angiosperms (AN-jee-oh-spermz) Plants that have seeds inside fruit.
anther (AN-thur) The part of the stamen on which pollen is found.
cells (SELZ) The basic parts of living things.
chlorophyll (KLOR-uh-fil) Green matter inside plants that allows them to use sunlight to make their own food.
conifers (KAH-nih-furz) Trees that have needlelike leaves and grow cones.
fertilization (fur-tuh-luh-ZAY-shun) Putting male cells inside an egg to make new plants.
fibrous root system (FY-brus ROOT SIS-tem) A root system that is usually formed by thin, branching roots that grow from the stem.
germination (jer-muh-NAY-shun) The way in which a seed or spore begins to sprout, or grow.
gymnosperms (JIM-noh-spermz) Plants that have seeds without fruit.
insects (IN-sekts) Small creatures that often have six legs and wings.
nutrients (NOO-tree-ints) Food that a living thing needs to live and grow.
ovary (OH-vuh-ree) The female part of a plant.
photosynthesis (foh-toh-SIN-thuh-sus) The way in which green plants make their own food from sunlight, water, and a gas called carbon dioxide.
pollination (pah-luh-NAY-shun) Spreading pollen from one plant to another so that the plants can reproduce.
reproduce (ree-pruh-DOOS) To make more of something.
spores (SPORZ) Special cells that can grow into new living things.
stamen (STAY-mun) The male part of a flower. The stamen carries the pollen.
stigma (STIG-muh) The sticky bump on a flower that catches pollen.
style (STYL) A long, tubelike part of a flower that carries pollen to the ovary.
taproots (TAP-roots) Straight roots that grow down and get narrower toward the bottom. Other roots may sprout from a taproot.

Index

A
angiosperms, 6, 14, 17–18, 21
anther, 21
asexual reproduction, 17

C
cells, 10, 17, 21
chlorophyll, 10
conifers, 6

F
fertilization, 17–18, 21
fibrous root system, 9
flower(s), 5–6, 17–18, 21

fruit(s), 5–6, 14, 21

G
germination, 14, 17
gymnosperms, 6, 14, 17–18

L
leaves, 5–6, 9–10, 13–14, 17

P
photosynthesis, 10, 13, 22

R
root(s), 5, 9–10, 14, 17

S
seeds, 5–6, 14, 17, 21–22
sexual reproduction, 17
spores, 17
stamen, 21
stem(s), 5, 9, 13–14, 17
stigma, 21
style, 21

T
taproots, 9

Web Sites

Due to the changing nature of Internet links, PowerKids Press has developed an online list of Web sites related to the subject of this book. This site is updated regularly. Please use this link to access the list:
www.powerkidslinks.com/gosci/plant/